CLIMATE CHANGE

---

40 WAYS TO WIN A PRO GLOBAL WARMING ARGUMENT

by Mischa Wu

CLADD
PUBLISHING

Cladd Publishing Inc.
USA

This publication is designed to provide accurate information regarding the subject matter covered. It is sold with the understanding that neither the author nor the publisher is providing medical, legal or other professional advice or services. Always seek advice from a competent professional before using any of the information in this book. The author and the publisher specifically disclaim any liability that is incurred from the use or application of the contents of this book.

Climate Change: 40 Ways To Win a Pro Global Warming Argument

ISBN 978-1-946881-17-5 (e-book)
ISBN 978-1-946881-18-2 (paperback)

CONTENTS

Global Warming & Climate Change Basics

It's Our Duty

Melting ice-caps, the destruction of vegetation and wildlife, along with violent surges of hurricanes, are all concerning reasons to understand how climate change can affect mankind. Global warming's massive impact on social, economic, and physical health is alarming!

It's our duty as citizens of the world, to be adopting a more responsible lifestyle. We do not need to wait for governments or scientist to find a solution. Each individual can be making a difference right now. It's the only practical way to save our planet, before it is too late.

GLOBAL WARMING VS. CLIMATE CHANGE

- **Global Warming**

Global warming refers to the constant increase of temperatures across the entire planet. However, the heating trend became obvious in the late 1970s, causing more organizations to get involved to find solutions. The main cause of the spike, was due to the increase in fossil fuels, brought on by the industrial revolution.

- **Climate Change**

Climate change refers to the various ways that our planet is reacting to heat-trapped gases in the atmosphere.

These reactions transpire into global warming, sea level rise, Greenland's loss in ice mass, Antarctica, the Arctic and mountain glaciers, shifts in flower blooming, and extreme weather patterns and storm surges.

5 KEY POINTS OF GLOBAL WARMING

1. **Rise in Sea Level:** Sea levels are rising in many areas of the world.
2. **Rise in Earth's Average Temperature:** Global temperature has steadily increased during the past 150 years.
3. **Rise in Ocean Temperatures:** The acceleration of vehicles and factories have resulted in an increase in greenhouse gases getting trapped in the atmosphere.
4. **Shrinking Glaciers:** The glaciers on several mountain ranges, particularly in Greenland and Antarctica, are retreating due to climate change.
5. **Ocean Acidification:** Acid levels in the oceans are increasing. This is due to emission of more harmful gases being absorbed. Oceans are one of our planets natural ways of

removing CO2 from the atmosphere. However, we are producing more than it can handle, causing an over acidification of our oceans.

CO2 IN THE ATMOSPHERE

WHY IT MATTERS: Carbon dioxide roughly makes up 80% of the United States greenhouse gas emissions. Fossil fuels and certain chemical reactions produce an odorless, colorless gas. This gas is responsible for trapping heat in the atmosphere. Despite our planets natural system of removing CO2, levels are the highest they have ever been.

DROUGHTS

WHY IT MATTERS: Droughts occur when there is a major imbalance between evaporation and precipitation. This in turn, causes a prolonged period of dry weather. Droughts have a devastating impact on our health, water and food supplies, animals, and the soil.

RISING SEA LEVELS (GLOBAL MEAN SEA LEVEL - GMSL)

WHY IT MATTERS: Our rising sea level is caused by melting ice and a warming ocean. Sea levels continue to rise, leaving coastal areas more vulnerable to flooding and storm surges. This also creates another issue; salt water is now steadily seeping into freshwater aquifers.

TEMPERATURE

WHY IT MATTERS: Temperature rise contributes to global climate change. It causes spikes in droughts, typhoons, hurricanes, wildfires, and habitat changes.

SEA SURFACE TEMPERATURE (SST)

WHY IT MATTERS: Our oceans absorb excess heat, causing them to get warmer. This issue affects marine life, decreases healthy fish populations, boost algal blooms, and kills coral. Higher sea surface temperatures increase atmospheric water vapor. SST drives a higher risk of extreme weather events, droughts, and strange storm patterns.

ARCTIC AND ANTARCTIC SEA ICE EXTENT

WHY IT MATTERS: The polar ice caps have existed for millions of years and are reliable proof of climate change. Polar Caps reflect sunlight providing the world with high albedo (reflectivity). This in turn, helps deflect solar radiation, dramatically cooling down our Earth.

NATURE CAN'T KEEP UP

NATURAL CARBON SEQUESTRATION

Natural carbon sequestration is the process where nature removes excess carbon dioxide out of our atmosphere. There are many sources of carbon like: Humans, animals, plants during the night, forest fires, volcanic eruptions, and magma reservoirs. With all of this carbon dioxide being put into the atmosphere, levels must be kept in balance otherwise the surface of the planet would quickly overheat.

Nature designed trees, oceans, soil and some animals to be effective carbon sponges. All organic life on this planet is carbon based. Thus, when plants and animals die, much of the carbon goes back into the ground where it has little impact on global warming.

Nature has a tremendously effective way of dealing with carbon dioxide. However, as the industrial revolution approached, and mankind frivolously used and disposed resources without the environment in mind, our planet become unable to handle the overload.

For example, our oceans have absorbed so much carbon dioxide, they are becoming saturated and acidic. Many tree planting programs have been initiated, but it's going to take a long time before they are mature enough to provide sequestration benefits.

Acting to reduce our carbon foot print on an individual daily basis, is just as important as acting on a long-term global scale.

40 Ways to Win a Pro Global Warming/Climate Change Argument

Whether you're in a debate group at school, or having a heated discussion with friends; these 40 answers to common objections will have them thinking twice about their stance on global warming.

Number 1: There Is No Real Evidence

OBJECTION: Despite what computer models indicates, there is no solid evidence of significant global warming.

ANSWER: Global warming is not only generated using a computer model. It is a conclusion derived through observations, measuring and many other scientific indicators. The most straightforward evidence is the actual global surface temperature records.

Number 2: One Hot Year Does Not Make Global Warming

OBJECTION: Records are set all the time. One warm year is not global warming.

ANSWER: A single year taken by itself, does not establish a warming trend. That is why computer models and records are accumulated over long periods of time.

Number 3: Temperature Records Are Unreliable

OBJECTION: Surface temperature records are full of assumptions and little to no verifiable facts.

ANSWER: It is true that weather station locations can make a difference in their surface temperature readings. However, scientist are focused on trends over long periods of time and across the entire globe.

NUMBER 4: ONE HUNDRED YEARS IS NOT ENOUGH DATA

OBJECTION: One hundred years of recorded global surface temperatures is not long enough to draw a conclusion of this magnitude.

ANSWER: Scientist use not only the last 150 years of recorded surface temperatures, but also a host of other important data to draw their conclusion. Another great source of temperature data is the borehole measurements. This involves drilling a deep hole and measuring the temperature of the earth at various depths.

NUMBER 5: GLACIERS HAVE ALWAYS GROWN AND RECEDED

OBJECTION: A few glaciers receding today is not proof of global warming.

ANSWER: Yes, glaciers have grown and receded over long periods of time. But today we are seeing this accelerating right in front of our eyes. Sea ice in the arctic is reaching new record declines and is continuing. Greenland's massive ice sheet has been losing nearly 100 gigatons of ice annually in recent years. Along with the thawing of ancient permafrost.

NUMBER 6: GLOBAL WARMING IS CAUSED BY THE URBAN ISLANDS EFFECT

OBJECTION: The rise of global average temperatures is an illusion based on urbanization of land around weather stations.

ANSWER: The urban Island Effect It is a real phenomenon, but scientists have taken steps to remove its influence from the raw data.

Number 7: Mauna Loa Is a Volcano

OBJECTION: CO_2 levels are recorded on top of Mauna Loa which is an active volcano! No wonder the levels are so extremely high.

ANSWER: Yes, it's true, Mauna Loa is an active volcano. By using scientific indicators like "wind direction," they are able to ensure that the readings are not contaminated by any out-gassing.

Number 8: Scientist Aren't Even Sure

OBJECTION: Even the scientists don't know if climate is changing more than normal, and if it's mankind's fault or the farting cows.

ANSWER: Scientists are always aware that new data is constantly evolving and expanding human knowledge. The Greenhouse-Gas theory is over 100 years old. The first predictions of anthropogenic global warming came in 1896. Time has only strengthened and helped refine those conclusions.

NUMBER 9: IT'S BEEN COLDER THAN NORMAL IN MY TOWN

OBJECTION: It was way colder than normal in my town, proving that there is no global warming.

ANSWER: Climate can never be drawn from a single data point, hot or cold. Scientist need a globalized system of measuring across long periods of time. Local temperatures don't mean much in their studies.

NUMBER 10: ANTARCTIC ICE IS GROWING

OBJECTION: The Antarctic ice sheets are growing, which wouldn't be happening if global warming were actually real.

ANSWER: Even if the Antarctic warmed by 20 degrees from -55 degrees Celsius. It would still leave it below freezing, so the snow wouldn't necessarily melt.

Number 11: Satellites Show Cooling

OBJECTION: Satellite readings show that the earth is cooling.

ANSWER: Satellite readings are very complicated. They must take into accounting the decay of the satellite orbits, splicing together records from numerous instruments, and trying to separate the signals of the atmosphere of origin. Satellite readings are not as simple as the thermometers that are hanging outside our homes.

Number 12: What About Mid-Century Cooling?

OBJECTION: There was global cooling in the '40s, '50s, and '60s.

ANSWER: Global warming is a large and complex system. Scientist use long term data to determine climate change. these 20 years are simply not enough data on their own to determine whether cooling was actually happening.

Number 13: Global warming stopped in 1998

OBJECTION: Global temperatures have been trending down since 1998. Global warming is over.

ANSWER: During 1998, the weather system known as El Nino was eerily strong. Choosing that year as a starting point, demonstrates why it is necessary to derive data from various sources around the world, over long periods of time.

Number 14: Antarctic Sea Ice Is Increasing

OBJECTION: Sea-Ice is shrinking in the Arctic, but growing in the Antarctic. This appears like a perfect balance.

ANSWER: In fact, it is completely in line with the computer models expectations. The model predicts that CO2 will have an extremely large effect in the north. The reasons for this is that the amount of land in the northern hemisphere is greater. In addition, the ocean's thermal inertia can delay any temperature signal from the ongoing absorption of heat. Lastly, the circumpolar current creates a natural buffer, preventing warm water from the tropics transporting heat to the South Pole. The north does not have this buffer.

NUMBER 15: CLIMATE SENSITIVITY IS LOW

OBJECTION: The small amount of temperature rise, should not send the entire world in hysteria.

ANSWER: The reason for this is primarily the large heat capacity of the oceans. This is commonly referred to as the climate system's thermal inertia. Our planets heat is increasing dramatically, but the oceans are absorbing much of it. Thus, creating so much danger for sea life, rise of sea levels, and increased storm systems.

NUMBER 16: SOME STATIONS SHOW COOLING

OBJECTION: Some stations show cooling trends. It should would be warming everywhere.

ANSWER: Global warming is based on long-term increases in global and seasonal average surface temperatures. Scientist will not use short-term temperature readings, based on specific locations to determine a global climate trend.

NUMBER 17: GLOBAL WARMING IS A HOAX

OBJECTION: Global warming is a hoax perpetrated by environmental extremists and liberals who want an excuse for more big government.

ANSWER: There are 13 world organizations that accept anthropogenic global warming as real and scientifically well-supported. We need our governments and global organizations to support a large-scale change for the better good of all mankind.

NUMBER 18: THERE IS NO CONSENSUS

OBJECTION: Climate is complicated and there are lots of competing theories and unsolved mysteries.

ANSWER: No one in the science community is debating whether changes in atmospheric CO_2 concentrations alter the greenhouse effect. Nor if the current warming trend is outside of average climate patterns. Nor if sea levels have risen over the last century.

NUMBER 19: POSITION STATEMENTS HIDE DEBATE

OBJECTION: All those institutional position statements are obscuring the truth; real debate is in the scientific journals.

ANSWER: The best way to find current knowledge is in scientific literature. You will find that the majority of all scientist in the climate field, have concluded without a doubt, that global warming is a real issue.

NUMBER 20: CONSENSUS IS COLLUSION

Objection: Fewer scientists dare speak out against the findings of the IPCC, thanks to the pressure to conform.

Answer: The results of the computer models are similar to the physical representations of the climate system. It is increasingly hard to dispute the models, historical and current data, surface temperature stations, satellite readings, and the various other measurements available.

NUMBER 21: PEISER REFUTED ORESKES

OBJECTION: Oreskes was refuted by Benny Peiser, who did the exact same survey and found very different results.

ANSWER: If anything, Peiser's effort strengthens Oreskes' finding of a widespread consensus.

NUMBER 22: CLIMATE MODELS ARE UNPROVEN

OBJECTION: Why should we trust a computer model that has never had a confirmed prediction?

ANSWER: The climate is an extremely complex system, and our models are designed to predict the longer-term effects of climate change. It is common knowledge that over time, we will begin experiencing many of the issues predicted by the computer models. Currently there have been many global temperature predictions that have been validated, and more will come.

The models have successfully predicted that surface warming should be accompanied by cooling of the stratosphere, and this has now been taking place. models have also predicted warming of the lower, mid, and upper troposphere. Models have predicted warming of ocean surface waters. Along with an energy imbalance between incoming sunlight and outgoing infrared radiation, to name a few.

NUMBER 23: MODELS DON'T ACCOUNT FOR CLOUDS

OBJECTION: The climate models don't even take cloud effects into account.

ANSWER: The atmospheric global climate models are designed using four different methods. One of the four models takes the effects of clouds into account.

NUMBER 23: AEROSOLS SHOULD MEAN MORE WARMING IN THE SOUTH

OBJECTION: Scientists claim that global warming from greenhouse gases is being countered somewhat by global dimming from aerosol pollution.

ANSWER: Aerosol cooling does affect the Northern Hemisphere, where most aerosols are created, more than the southern hemisphere. The problem starts by assuming that CO_2 concentrations imply that our planet should be heating up evenly.

This phenomenon was predicted by the model, that CO2-dominated warming will disproportionately affect the north. Another important piece of the puzzle, is how the ocean dynamics in the north and south are different. More heat mixes into deeper waters of the Southern Ocean.

NUMBER 24: WE CAN'T EVEN PREDICT THE WEATHER NEXT WEEK

OBJECTION: Scientists can't even predict the weather next week.

ANSWER: Climate and weather are very different things. Scientist are not studying daily weather, they are observing and measuring climate patterns over hundreds/thousands of years. They are not concerned with whether we are going to have a sunny weekend or if it will be pouring rain. The study of our climate is a great indicator of how our consumption and general behavior is effecting our planet.

Number 25: Chaotic Systems Are Not Predictable

OBJECTION: Climate is a chaotic system, and it cannot be predicted.

ANSWER: Our planets climate is really not chaotic. In fact, there are strong gradual patterns that are identifiable. The confusion starts with the mix up between local daily weather and long term global climate. Once this is straightened out, the rest is obvious.

Number 26: Hansen Has Been Wrong Before

OBJECTION: In 1988, Hansen predicted dire warming over the next decade.

ANSWER: Here is what actually happens:
In 1988, James Hansen went before the Senate on the danger of global warming. During that testimony, he presented a graph. His graph represented three scenarios of future emissions and volcanism.

a) Line A was a temperature trend based on rapid emissions, with no large volcanic event.
b) Line B was based on modest emissions, and one large volcanic eruption in the mid-1990s.
c) Line C was similar to line B, and included the same volcanic eruption, but showed reductions in the growth of CO_2 emission by the turn of the century. The result of hypothetical government controls on CO_2.

Line B came very close to predicting what happened. Hansen was right, the model proved successful.

NUMBER 27: THEY PREDICTED GLOBAL COOLING IN THE 70'S

OBJECTION: The alarmists were predicting the onset of an ice age in the '70s.

ANSWER: In the 1970s, there was a small book, a few articles in popular magazines, and a minute amount of scientific speculation of an ice-age. Today, you have a widespread scientific consensus, supported by every major scientific institution, confirming that the temperature is rising.

NUMBER 28: GREENLAND USED TO BE GREEN

OBJECTION: When the Vikings settled it, Greenland was lush green.

ANSWER: Greenland is part of a single region. It cannot be used to represent a global climate shift.

NUMBER 29: THE HOCKEY STICK MODEL IS BROKEN

OBJECTION: Scientist own climate data models broke the hockey stick pattern.

ANSWER: Although each of the temperature models are different, they all show some similar patterns of temperature change over the last several centuries. Most interesting is the fact that each record shows that the 20th century is the warmest on record, and that warming was most dramatic after 1920.
It also concluded that current northern hemisphere surface temperatures are significantly higher than during the peak of the Medieval Warm Period

NUMBER 30: NEWFOUNDLAND WAS FULL OF GRAPES

OBJECTION: Newfoundland was so warm in the Medieval Warm Period that Vikings brought grapes back to Europe.

ANSWER: Once again: you can't draw conclusions about global climate from a single region, or even a few. You need detailed long-term analysis of climates from around the world.

NUMBER 31: GLOBAL WARMING IS PART OF A NATURAL CYCLE

OBJECTION: Current warming is just part of a natural cycle.

ANSWER: It is true that there are natural cycles and variations in global climate. However, there has been no natural causes identified for the extreme climate cycles we are witnessing today.

NUMBER 32: MARS AND PLUTO ARE WARMING TOO

OBJECTION: Global warming is happening on Mars and Pluto.

ANSWER: The only common factor that Earth and Mars share is the sun. Other than that, this objection is fact deficient.

NUMBER 33: NATURAL EMISSIONS DWARF HUMAN EMISSIONS

OBJECTION: 150 billion tonnes of carbon go into the atmosphere from natural processes annually. This is roughly 30 times the amount of carbon humans emit.

ANSWER: Natural emissions have always been in balance. Even though the carbon being emitted by nature is much larger than our own, we have now tipped the balance of its give and take. We put approximately 6 gigatonnes of carbon into the air, but unlike nature, we are not taking any out. When we began burning fossil fuels over 150 years ago, the atmospheric concentration was relatively stable for the previous several thousand years. Today it has now risen by over 35%.

NUMBER 34: THE CO2 RISE IS NATURAL

OBJECTION: It's clear from ice cores and other geological history that CO2 fluctuates naturally.

ANSWER: Yes, CO2 does fluctuate naturally. However, we have been on a steady increase with no foreseeable decline. We emit billions of tons of CO2 into the air, and now there is more CO2. It's pretty straightforward.

NUMBER 35: WE ARE JUST RECOVERING FROM THE ICE AGE

OBJECTION: Today's warming is just a recovery from the Little Ice Age.

ANSWER: This idea assumes that there is a baseline climate that the planet eventually returns to. But even if some recovery was expected from that specific point, why have we now exceeded it with no return in site?

NUMBER 36: IT'S THE SUN

OBJECTION: The sun is the source of warmth on earth, any globalized warming is likely due to changes in solar radiation.

ANSWER: According to the World Radiation Center there has been no increase in solar irradiance since at least 1978, when satellite observations began.

NUMBER 37: IT'S THE WATER VAPORS

OBJECTION: H_2O accounts for 95% of the greenhouse effect; CO_2 is not significant.

ANSWER: CO_2 contributes anywhere from 9% to 30% to the overall greenhouse effect. The 95% number was not produced through a reliable scientific source.

NUMBER 38: WHAT'S WRONG WITH WARMER WEATHER

OBJECTION: The earth has had much warmer climates in the past. What's so special about the current climate?

ANSWER: The real issue is not what the temperature is now, or will be. The problem is how fast it is moving. Rapid change is extremely dangerous for sustaining life.

The negative impacts of global warming would be disastrous to agriculture, health, economy and environment.

NUMBER 39: THE CLIMATE IS ALWAYS CHANGING

OBJECTION: Climate has always changed. Why are we worried now?

ANSWER: Yes, climate has varied in the past, for many different reasons. Present-day climate change is well understood, and different. Rapid warming on a global scale is rare in the geological records. There is strong evidence that whenever an extreme change has occured, whatever the cause, it was a catastrophic event for all living things.

NUMBER 40: THE MEDIEVAL WARM PERIOD WAS JUST AS WARM AS IT IS TODAY

OBJECTION: It was just as warm in the Medieval Warm Period (MWP) as it is today.

ANSWER: There is no good evidence that the MWP was a globally warm period comparable to today. This is more of a myth than anything else.

SCHOOL YOUR SKEPTICS WITH THESE 74 ONE-LINERS

School those uninformed global warming skeptics, with these clever one-liners. Use them to make believers out of everyone!

SKEPTIC: "CLIMATE HAS CHANGED BEFORE"

1. Climate reacts to whatever forced it to change at that time; humans are now the dominant force today.

SKEPTIC: "IT'S THE SUN"

2.   In the last 35 years of global warming the World Radiation Center has recorded no increase in solar irradiance.

SKEPTIC: "WARMING TEMPERATURES ARE NOT BAD"

3.   The droughts, rising sea level and severe storms will be disastrous to agriculture, health and environment.

SKEPTIC: "THERE IS NO CONSENSUS"

4.   97% of climate experts agree humans are causing global warming.

SKEPTIC: "IT'S COOLING"

5.   2000-2009 was the hottest spell in recorded history, and more records are being broke all the time.

Skeptic: "Models are unreliable"

6. Models have successfully reproduced temperatures since 1900 globally, by land, in the air and the ocean.

Skeptic: "Temperature record is unreliable"

7. The warming trend is similar for rural and urban areas, which is measured both by thermometers and satellites.

Skeptic: "Animals and plants can adapt"

8. Global climate change has already caused mass extinctions of species that cannot adapt on short time scales.

Skeptic: "It hasn't warmed since 1998"

9. Earth's climate has continued warming since 1998, shattering all previous records.

Skeptic: "Antarctica is Gaining Ice"

10. Satellites show Antarctica losing ice at a faster rate than previous recorded history.

Skeptic: "Ice Age Predicted in the 70s"

11. The majority of climate models in the 1970s-predicted warming, not cooling.

Skeptic: "CO2 Lags Temperature"

12. CO2 didn't trigger warming from the last ice ages, but it did amplify the effects.

Skeptic: "Climate Sensitivity is Low"

13. Positive evaluation is now confirmed by many different forms of evidence.

Skeptic: "We're Heading Into an Ice Age"

14. All knowledge leads scientist to only one conclusion, our planet has been continually warming at alarming rates.

SKEPTIC: "OCEAN ACIDIFICATION ISN'T SERIOUS"

15. Ocean acidification threatens entire marine food chains.

SKEPTIC: "HOCKEY STICK IS BROKEN"

16. Current studies agree that the current increased global temperatures are the highest in the last 1000 years.

SKEPTIC: "CLIMATEGATE CRU EMAILS SUGGEST CONSPIRACY"

17. Investigations have cleared scientists of any violations in the media-hyped email scandal.

SKEPTIC: "HURRICANES AREN'T LINKED TO GLOBAL WARMING"

18. Evidence shows that hurricanes are getting stronger due to global warming.

SKEPTIC: "AL GORE GOT IT WRONG"

19. Al Gore's book is quite accurate, and far more accurate than contrarian books.

SKEPTIC: "GLACIERS ARE GROWING"

20. Most glaciers are retreating, creating a dire problem for the millions who rely on glaciers for water.

SKEPTIC: "COSMIC RAYS ARE TO BLAME"

21. Cosmic rays show no increase over the last 30 years, and have had little impact on global warming.

SKEPTIC: "1934 - HOTTEST YEAR ON RECORD"

22. 1934 was the hottest year in the US, but not globally.

SKEPTIC: "IT'S EXTREMELY COLD IN MY TOWN!"

23. A local cold day has nothing to do with the long-term trend of increasing global temperatures.

SKEPTIC: "EXTREME WEATHER ISN'T CAUSED BY GLOBAL WARMING"

24. Extreme weather is being worsened by global warming.

SKEPTIC: "SEA LEVEL RISE IS EXAGGERATED"

25. A plethora of different scientific measurements show a steady rise in sea levels across the board.

SKEPTIC: "IT'S URBAN HEAT ISLAND EFFECT"

26. Urban and rural regions show the same warming trend, as the rest of the world based on long term data.

SKEPTIC: "MEDIEVAL WARM PERIOD WAS WARMER"

27. The planets average temperature is now higher than in medieval times.

SKEPTIC: "MARS IS WARMING"

28. Scientist do not have enough long-term data to prove Mars is actually warming.

SKEPTIC: "ARCTIC ICE-MELT IS A NATURAL CYCLE"

29. Thick Arctic sea ice is undergoing a rapid retreat based on studies, measurements and data.

SKEPTIC: "INCREASING CO2 HAS LITTLE EFFECT"

30. The CO2 effect has been observed by many different measurements, and all conclude its powerful influence.

SKEPTIC: "OCEANS ARE COOLING"

31. The most recent ocean measurements show consistent warming.

SKEPTIC: "IT'S A 1500-YEAR CYCLE"

32. Ancient natural cycles are based on a balanced system, whereas today we are dealing with mankind's footprint.

SKEPTIC: "HUMAN CO2 IS A TINY % OF EMISSIONS"

33. The natural cycle adds and removes CO2, humans only add extra CO2 without removing any.

SKEPTIC: "IPCC IS ALARMIST"

34. Numerous sources have documented how IPCC predictions are likely to underestimate the climate response, not the opposite.

SKEPTIC: "WATER VAPOR IS THE MOST POWERFUL GREENHOUSE GAS"

35. Sky-rocketing CO2 increases atmospheric water vapor, which makes global warming even worse.

SKEPTIC: "POLAR BEAR NUMBERS ARE INCREASING"

36. Polar bears are in danger of extinction as well as many other species.

SKEPTIC: "CO2 LIMITS WILL HARM THE ECONOMY"

37. If we don't limit our CO2 output today, then our children will pay the price for our ignorance tomorrow.

SKEPTIC: "IT'S NOT HAPPENING"

38. There are many forms of evidence proving global warming is an absolute.

SKEPTIC: "GREENLAND WAS GREEN"

39. Other parts of the planet got colder when Greenland became warmer.

SKEPTIC: "GREENLAND IS GAINING ICE"

40. Greenland is losing ice, as confirmed by satellite measurements.

SKEPTIC: "CO2 IS NOT A POLLUTANT"

41. Because of its impacts on the climate, CO2 presents a danger to public health and welfare, thus categorizing it as an air pollutant.

SKEPTIC: "THERE'S NO EMPIRICAL EVIDENCE"

42. Models, measurements and raw data shows the increase in global warming directly correlates to humans.

SKEPTIC: "CO2 IS PLANT FOOD"

43. The effects of enhanced CO2 on terrestrial plants are variable and complex and dependent on numerous factors.

SKEPTIC: "OTHER PLANETS ARE WARMING"

44. There is no absolute proof that Mars and Jupiter are warming.

SKEPTIC: "ARCTIC SEA ICE HAS RECOVERED"

45. Thick arctic sea ice is in rapid retreat, based on scientific measurements and data.

SKEPTIC: "THERE'S NO CORRELATION BETWEEN CO2 AND TEMPERATURE"

46. There is long-term connections between CO2 and global temperature; other effects are only short-term.

SKEPTIC: "WE'RE COMING OUT OF THE LITTLE ICE AGE"

47. Scientists have determined that current global warming is not simply a warming phase, meant to reverse the effects of a "little ice age".

SKEPTIC: "IT COOLED MID-CENTURY"

48. Mid-century cooling involved increased aerosol levels, and is irrelevant for recent global warming.

SKEPTIC: "CO2 WAS HIGHER IN THE PAST"

49. When CO2 was higher in the past, it was due to the sun being cooler.

SKEPTIC: "IT WARMED BEFORE 1940 WHEN CO2 WAS LOW"

50. 20th century warming was due to several issues, including rising CO2 levels.

SKEPTIC: "SATELLITES SHOW NO WARMING IN THE TROPOSPHERE"

51. The most recent satellite data show that the entire planet is warming.

SKEPTIC: "IT'S AEROSOLS"

52. Aerosols have been masking global warming, that is why it is worse than initially thought.

SKEPTIC: "2009-2010 WINTER SAW RECORD COLD SPELLS"

53. An extreme cold day in winter at a specific location, has nothing to do with the trend of the world.

SKEPTIC: "IT'S EL NIÑO"

54. El Nino is not responsible for the trend of global warming, but could be worsened by it.

Skeptic: "Mt. Kilimanjaro's ice loss is due to land use"

55. Most glaciers are quickly retreating worldwide, but a few are experiencing a change based on human settlements.

Skeptic: "It's not us"

56. Multiple sets of independent observations find a human footprint on climate change.

Skeptic: "It's a natural cycle"

57. Scientific data shows that natural cycles are smooth, but today we have seen an extreme spike coinciding with humans and their output.

Skeptic: "There's no tropospheric hot spot"

58. There is a clear short-term hot spot, and evidence is mounting for the discovery of a long-term hot spot.

SKEPTIC: "IT'S PACIFIC DECADAL OSCILLATION"

59. The PDO shows no trend, and therefore it is not responsible for the increase of global warming.

SKEPTIC: "SCIENTISTS CAN'T EVEN PREDICT WEATHER"

60. Weather and climate are different; climate predictions are based on long-term weather data across the entire world.

SKEPTIC: "2ND LAW OF THERMODYNAMICS CONTRADICTS GREENHOUSE THEORY"

61. The 2nd law of thermodynamics is consistent with the greenhouse theory.

SKEPTIC: "IT'S THE OCEAN"

62. The oceans are warming, thus becoming more acidic and threatening the food chain.

Skeptic: "Volcanoes emit more CO2 than humans"

63. Humans emit 100 times more on an average than CO2 from volcanoes.

Skeptic: "It's methane"

64. Methane does have a slight effect on global warming, but could get worse when the permafrost begins to melt.

Skeptic: "Record snowfall disproves global warming"

65. Warming leads to increased evaporation and precipitation, which would increase snow during the winter.

Skeptic: "Peer review process was corrupted"

66. Numerous Independent Reviews and studies have concluded, that global warming is a real threat.

SKEPTIC: "RENEWABLE ENERGY IS TOO EXPENSIVE"

67. When people demand renewable energy sources, more companies will come into the market, creating competitive prices.

SKEPTIC: "PHIL JONES SAYS NO GLOBAL WARMING SINCE 1995"

68. Phil Jones was misquoted by skeptics for their own purpose.

SKEPTIC: "IT'S NOT URGENT"

69. A large portion of warming is delayed, we could already be past the tipping points.

SKEPTIC: "IT'S OZONE"

70. Ozone has only a small effect on global warming based on the data.

SKEPTIC: "FREEDOM OF INFORMATION (FOI) REQUESTS WERE IGNORED"

71. CRU was found not guilty of ignoring the FOI request.

SKEPTIC: "DMI SHOWS A COOLING ARCTIC"

72. The annual average Arctic temperatures have risen sharply in recent years.

SKEPTIC: "IT'S SATELLITE MICROWAVE TRANSMISSIONS"

73. Satellite transmissions are extremely small and irrelevant to the larger scope of global warming.

SKEPTIC: "WE DIDN'T HAVE GLOBAL WARMING DURING THE INDUSTRIAL REVOLUTION"

74. CO_2 emissions were much smaller 100 years ago, today factories exist in every town across the globe.

33 Easy Ways to Reduce Global Warming Today

Here are 33 easy ways that you can do your part in reducing global warming.

1. **Replace Regular Incandescent Light bulb:**
    Replace your regular incandescent light bulb with compact fluorescent light (CFL) bulbs. They consume 70% less energy and have a longer lifespan.

2. Drive Less or Carpool:

By driving less, you are saving fuel, and helping reduce the effects of global warming. The largest sources of pollution are caused by oil and gasoline. Cutting down consumption, is a huge step to reducing waste.

ALSO TRY:

- Carpooling
- Moving closer to work
- Finding a job closer to home
- Walk, jog, skateboard, scooter or rollerblades and skates

3. Reduce, Reuse, Recycle:

REDUCE your need to buy new products or simply use less. This will result in a smaller amount of waste. Consider buying eco-friendly products.

REUSE bottles, plastic containers, glass jars and other items. Reusing water bottles, yogurt cups, bread ties, and other items is being eco-conscious. It will lessen what goes into the landfills, and the amount factories will produce.

RECYCLING unwanted paper, bottles, bags, and more. This is an easy way to make a huge difference. If possible, upcycle tables, furniture, and other outdated items to keep landfills clean. You can recycle almost anything like: paper, aluminum foils, cans, and newspapers.

4. Go Solar:

Having solar panels installed is becoming more affordable and readily available. Incentives and discounts given by government and energy companies make solar energy something to consider. Solar panels provide energy when the energy companies cannot. It saves you money and gives you peace-of-mind.

5. Reusable Shopping Bags:

Using your own reusable shopping bag, makes a huge impact on the reduction of plastics. Reducing plastics in our everyday life is a very important.

6. **Buy Energy-Efficient Appliances:**
   Energy-efficient products can help you to save energy, money and reduce your carbon footprint.

7. **Reduce Waste:**
   Landfills are a major contributor of methane and other greenhouse gases. When the waste is burnt, it releases toxic gases in the atmosphere which greatly contribute to global warming. Reusing and recycling old items will significantly reduce the toxicity of our landfills.

8. **Use Less Hot Water:**
   Buy energy saving geysers and dishwasher for your home. Avoid washing clothes in hot water. Also, avoid taking frequent showers and use less hot water when possible.

9. **Avoid Products with a Lot of Packaging:**
    Purchase fresh produce, products with less packaging or items in bulk. This will decrease the amount of plastics from packages in our landfills.

10. **Install a Programmable Thermostat:**
    A programmable thermostat is becoming very affordable if you do not already have one. Lower your thermostat 2 degrees in the winter. Instead of making your home a hot sauna, try putting on extra layers. You can also do this in the summer time to save more energy.

11. **Turn Off the Lights:**
    If you're not using a room, there's no need for the light to be on. Leaving lights on is very wasteful.

12. **Turn off Electronic Devices:**
    Turn off electronic devices when you are not using them. Unnecessary usage is wasteful.

13. Plant a Tree:

Planting trees can help in reducing global warming, and cleans the air in close proximity. They not only give oxygen, but also take in carbon dioxide.

14. Use Clean Fuel:

Support companies that are producing eco-friendly choices.

15. Look for Better Options:

If you can't afford an electric car, buy a vehicle that uses the least amount of gasoline possible.

16. Save Energy:

When you consume less, the less carbon dioxide is released into the atmosphere. Setting your thermostat, or changing the type of light bulb you use is a great start. But don't forget about the practical ways you can make a difference every day.

17. Replace Filters on Air Conditioner and Furnace:

If you haven't changed your air filter, you are wasting energy and breathing dirty air. Cleaning a dirty air filter can save several pounds of carbon dioxide a year. It can also reduce allergies, and asthma.

18. Go Green:

Using energy star appliances will save you money, and reduces energy wasted in your home.

19. Tune Your Car Regularly:

Regular vehicle maintenance will help your car function properly and emit less carbon dioxide.

20. Conserve Water:

If we added up the water wasted by brushing our teeth, we could give clean water to more than 23 nations with contaminated drinking water.

- By taking a quick shower instead of baths you will use 25% less water over the course of a year. That's hundreds of gallons saved!

21. Stop Idling Your Car:
Only idle your car if necessary. An average person idles for 15 minutes or more, and longer during colder months.

22. Eat Less Hamburger:
Cows emit methane into the air, which contributes to global warning. By lowering your (cow) meat intake, it will help dwindle the need for large cattle farms.

23. Use Clothesline to Dry Your Clothes:
Most clothes shouldn't be put in the dryer anyway, so this is a great tip for your fabric and the planet.

24. Eat Naturally:
Eating naturally is the best for your health. However, it also cuts down the energy used by factories who produce processed food, along with the plastics used for packaging.

25. Use a Kitchen Cloth Instead of Paper Towels:

Paper towels produce a lot of wasted energy, factory pollution, as well as large tree consumptions.

26. Check Your Tires:

Make sure your tires are inflated properly. If they are a little on the flat side, your vehicle is consuming more fuel which releases more CO_2 in the atmosphere.

27. Take Lunch in a Tupperware:

Each time you throw away a paper sack or plastic bag, more is being produced in a factory for your next purchase.

28. Wrap your water heater in insulation:

Wrap your water heater in insulation to save tons of energy. This not only helps the earth, but also your pocketbook.

29. Home Energy Audit:

Call a home energy audit company to identify areas that are not energy efficient. They will also help you discover ways that you can be more energy efficient, based on your unique household needs.

30. Become Part of the Global Warming Community:

Connecting with others who share the same concerns, helps strengthen the need for all of us to lessen our carbon footprint.

31. Celebrate Arbor Day and Earth day:

It's very important that you not only acknowledge the importance of these days, but also participate. Plant a tree, pick up trash, or join a forum to make a difference.

32. Become Aware of Your Contribution:

Make it your mission to learn more information about protecting the environment.

33. Spread the Awareness:

Always try your best to inform people about global warming. If we just keep this vital information to ourselves, then all of us will suffer the great consequences.

OUR MISSION

By being more mindful, we all can play our part in combating global warming. These easy tips will help preserve the planet for future generations.

We do not need to wait for governments, organizations, friends, family or even neighbors to make a change. It all starts with you, right now!

Made in the USA
Middletown, DE
01 July 2017